UAV PILOT
LOGBOOK

HOLDER'S NAME:

_____

Logbook number _____

Entries from _____

Through _____

Copyright © 2016 by PARHELION AEROSPACE GMBH

www.parhelionaerospace.com
Author: Michael L. Rampey
ISBN: 978-3-033-05808-8

HOLDER

Holder's Operator Certificate	Holder's Address
Certificate	
Number	
Date / place of initial qualification	
(Space for address change)	*(Space for address change)*

3

Notes on the recording of flight time in this logbook:

<u>By columns across the top of each page / per flight:</u>

Date
> Enter the current year in the space provided at the upper left-hand corner of each flight data page. Enter the day and month on which the flight commences in the spaces given in this column.

Aircraft
> Enter the make, model and aviation authority registration number of the aircraft flown.

Flight Location
> Enter the take-off location and time and the landing location and time. Note that the takeoff and landing may occur at the same place and need be entered only once for both phases of flight. Remark whether you use Local time or UTC.

Flight Duration
> Enter the duration of the flight, in hours and minutes or hours and tenths, in the column that corresponds to the type of aircraft being flown and again in the column marked 'Total'. In this manner it is possible to record separately the pilot's experience with Fixed-Wing, Rotary and Lighter-Than-Air (LTA) UAV aircraft and to sum the total flight time.

Flight Function
> Enter the duration of the flight, in hours and minutes or hours and tenths, in the column that corresponds to the role performed by pilot during the flight. In this manner it is possible to record the total amount of training received or instruction given separately from the total flight time. Cross-out the inappropriate function (e.g. either Training or Instructor).

Ld
> Enter the number of landings performed during the flight in this section.

Remarks / Endorsements
> Enter pertinent observations about the flight (e.g. observer name, incidents occurrence, weather conditions affecting flight, etc), as well as instructor/examiner signature when appropriate, in this section.

Notes on the recording of flight time in this logbook, continued:

Totals across the bottom of each page:

Totals this page:
Sum the durations entries from the columns above and enter these data in the spaces provided.

Totals brought forward:
Enter the '**Totals to date**' data from the *previous page* in the spaces provided.

Totals to Date:
Sum the 'Totals this page' and the 'Totals brought forward' entries and enter these data into the spaces provided. Note that the resultant '**Totals to date**' for Fixed Wing plus Rotary plus LTA should equal the entry for 'Total' time (bold-bordered box). The '**Totals to date**' for 'OIC/Solo' plus 'Training/Instructor' should equal the entry for 'Total' time as well.

Sign the page in the lower-right corner in the space provided to attest to the fact that the entries are true and accurate.

An example is provided on the next page.

Date	Aircraft	Location		Flight Duration				Function		LD	Remarks and Endorsements
Yr 2018	Make/Model	Takeoff	Landing	Fixed-wing	Rotary	LTA	Total	OIC/Solo	Student/Instructor	No.	
Dd/ Mm	Registration	Time L	Time L								
04|05	DJI P3P	Fly Hi, Parker Co.			0:30		0:30	0:30		1	Practice using waypoints
	N12345	8:00	8:30								
07|05	my PA X-1	47.1497° N, 7.2389° E		0:35			0:35	0:35		1	Obs: Geo. Washington
	N54321	12:10	12:45								

Example page:

In the example above the first two entries of a page and separately, below them, the corresponding totals, are shown as if the entire page had been filled-out and eight flights flown and entered.

The pilot has flown both Fixed-wing and Rotary aircraft and has logged the times separately. The duration of each flight has also been entered into the 'Total' column, so that the combined time of all aircraft flown can be calculated.

This pilot has entered the 'Totals this page' in the spaces provided at the bottom and has copied the corresponding 'Totals to date' from the previous page and entered them here as 'Totals brought forward.' The Fixed-wing time flown to date is therefore shown to be 7 hours 45 minutes, the total Rotary time to date is 5 hours 55 minutes and the pilot's absolute total time is 13 hours 40 minutes for all aircraft together. Note that this pilot has flown 1 hour 55 minutes while receiving instruction and 11 hours 45 minutes as OIC, which also sum to 13 hours 40 minutes.

	Fixed-wing	Rotary		Total	OIC/Solo	Student/Instructor	No.	
Totals this page:	4:15	3:40		7:55	6:00	1:55	8	*I certify that the entries in this log are true.*
Totals brought forward:	3:30	2:15		5:45	5:45	0	8	(Signature)
Totals to date:	7:45	5:55		13:40	11:45	1:55	16	

6

DATE	AIRCRAFT	LOCATION		FLIGHT DURATION				FUNCTION		LD	REMARKS AND ENDORSEMENTS
Yr	Make/Model	Takeoff	Landing	Fixed-wing	Rotary	LTA	Total	OIC/Solo	Student/Instructor	No.	
Dd/ Mm	Registration	Time	Time								
__/__											
__/__											
__/__											
__/__											
__/__											
__/__											
__/__											
__/__											
Totals this page:											I certify that the entries in this log are true.
Totals brought forward:											
Totals to date:											

7

DATE	AIRCRAFT	LOCATION		FLIGHT DURATION				FUNCTION		LD	REMARKS AND ENDORSEMENTS
Yr	Make/Model	Takeoff	Landing	Fixed-wing	Rotary	LTA	Total	OIC/Solo	Student/Instructor	No.	
Dd/ Mm	Registration	Time	Time								
__/__											
__/__											
__/__											
__/__											
__/__											
__/__											
__/__											
__/__											
Totals this page:											I certify that the entries in this log are true.
Totals brought forward:											
Totals to date:											

DATE	AIRCRAFT	LOCATION		FLIGHT DURATION				FUNCTION		LD	REMARKS AND ENDORSEMENTS
Yr	Make/Model	Takeoff	Landing	Fixed-wing	Rotary	LTA	Total	OIC/Solo	Student/ Instructor	No.	
Dd/ Mm	Registration	Time	Time								
__/__											
__/__											
__/__											
__/__											
__/__											
__/__											
__/__											
__/__											
Totals this page:											I certify that the entries in this log are true.
Totals brought forward:											
Totals to date:											

9

DATE	AIRCRAFT	LOCATION		FLIGHT DURATION				FUNCTION		LD	REMARKS AND ENDORSEMENTS
Yr	Make/Model	Takeoff	Landing	Fixed-wing	Rotary	LTA	Total	OIC/Solo	Student/ Instructor	No.	
Dd/ Mm	Registration	Time	Time								
__/__											
__/__											
__/__											
__/__											
__/__											
__/__											
__/__											
__/__											
Totals this page:											*I certify that the entries in this log are true.*
Totals brought forward:											
Totals to date:											

10

Date	Aircraft	Location		Flight Duration				Function		Ld	Remarks and Endorsements
Yr	Make/Model	Takeoff	Landing	Fixed-wing	Rotary	LTA	Total	OIC/Solo	Student/Instructor	No.	
Dd/ Mm	Registration	Time	Time								
__/__											
__/__											
__/__											
__/__											
__/__											
__/__											
__/__											
__/__											
Totals this page:											I certify that the entries in this log are true.
Totals brought forward:											
Totals to date:											

11

DATE	AIRCRAFT	LOCATION		FLIGHT DURATION				FUNCTION		LD	REMARKS AND ENDORSEMENTS
Yr	Make/Model	Takeoff	Landing	Fixed-wing	Rotary	LTA	Total	OIC/Solo	Student/Instructor	No.	
Dd/ Mm	Registration	Time	Time								
__ / __											
__ / __											
__ / __											
__ / __											
__ / __											
__ / __											
__ / __											
__ / __											
	Totals this page:										*I certify that the entries in this log are true.*
	Totals brought forward:										
	Totals to date:										

DATE	AIRCRAFT	LOCATION		FLIGHT DURATION				FUNCTION		LD	REMARKS AND ENDORSEMENTS
Yr	Make/Model	Takeoff	Landing	Fixed-wing	Rotary	LTA	Total	OIC/Solo	Student/ Instructor	No.	
Dd/ Mm	Registration	Time	Time								
__/__											
__/__											
__/__											
__/__											
__/__											
__/__											
__/__											
__/__											
Totals this page:											I certify that the entries in this log are true.
Totals brought forward:											
Totals to date:											

13

DATE	AIRCRAFT	LOCATION		FLIGHT DURATION				FUNCTION		LD	REMARKS AND ENDORSEMENTS
Yr	Make/Model	Takeoff	Landing	Fixed-wing	Rotary	LTA	Total	OIC/Solo	Student/Instructor	No.	
Dd/ Mm	Registration	Time	Time								
__/__											
__/__											
__/__											
__/__											
__/__											
__/__											
__/__											
__/__											
	Totals this page:										I certify that the entries in this log are true.
	Totals brought forward:										
	Totals to date:										

14

DATE	AIRCRAFT	LOCATION		FLIGHT DURATION				FUNCTION		LD	REMARKS AND ENDORSEMENTS
Yr	Make/Model	Takeoff	Landing	Fixed-wing	Rotary	LTA	Total	OIC/Solo	Student/ Instructor	No.	
Dd/ Mm	Registration	Time	Time								
__/__											
__/__											
__/__											
__/__											
__/__											
__/__											
__/__											
__/__											
Totals this page:											*I certify that the entries in this log are true.*
Totals brought forward:											
Totals to date:											

15

DATE	AIRCRAFT	LOCATION		FLIGHT DURATION				FUNCTION		LD	REMARKS AND ENDORSEMENTS
Yr	Make/Model	Takeoff	Landing	Fixed-wing	Rotary	LTA	Total	OIC/Solo	Student/ Instructor	No.	
Dd/ Mm	Registration	Time	Time								
__/__											
__/__											
__/__											
__/__											
__/__											
__/__											
__/__											
__/__											
		Totals this page:									I certify that the entries in this log are true.
		Totals brought forward:									
		Totals to date:									

16

DATE	AIRCRAFT	LOCATION		FLIGHT DURATION				FUNCTION		LD	REMARKS AND ENDORSEMENTS
Yr	Make/Model	Takeoff	Landing	Fixed-wing	Rotary	LTA	Total	OIC/Solo	Student/Instructor	No.	
Dd/ Mm	Registration	Time	Time								
__/__											
__/__											
__/__											
__/__											
__/__											
__/__											
__/__											
__/__											
Totals this page:											I certify that the entries in this log are true.
Totals brought forward:											
Totals to date:											

DATE	AIRCRAFT	LOCATION		FLIGHT DURATION				FUNCTION		LD	REMARKS AND ENDORSEMENTS
Yr	Make/Model	Takeoff	Landing	Fixed-wing	Rotary	LTA	Total	OIC/Solo	Student/Instructor	No.	
Dd/ Mm	Registration	Time	Time								
__/__											
__/__											
__/__											
__/__											
__/__											
__/__											
__/__											
__/__											
Totals this page:											*I certify that the entries in this log are true.*
Totals brought forward:											
Totals to date:											

Date	Aircraft		Location		Flight Duration				Function		LD	Remarks and Endorsements
Yr	Make/Model		Takeoff	Landing	Fixed-wing	Rotary	LTA	Total	OIC/Solo	Student/ Instructor	No.	
Dd/ Mm	Registration		Time	Time								
__/__												
__/__												
__/__												
__/__												
__/__												
__/__												
__/__												
__/__												
	Totals this page:											I certify that the entries in this log are true.
	Totals brought forward:											
	Totals to date:											

DATE	AIRCRAFT	LOCATION		FLIGHT DURATION				FUNCTION		LD	REMARKS AND ENDORSEMENTS
Yr	Make/Model	Takeoff	Landing	Fixed-wing	Rotary	LTA	Total	OIC/Solo	Student/ Instructor	No.	
Dd/ Mm	Registration	Time	Time								
__/__											
__/__											
__/__											
__/__											
__/__											
__/__											
__/__											
__/__											
Totals this page:											I certify that the entries in this log are true.
Totals brought forward:											
Totals to date:											

DATE	AIRCRAFT	LOCATION		FLIGHT DURATION				FUNCTION		LD	REMARKS AND ENDORSEMENTS
Yr	Make/Model	Takeoff	Landing	Fixed-wing	Rotary	LTA	Total	OIC/Solo	Student/ Instructor	No.	
Dd/ Mm	Registration	Time	Time								
__/__											
__/__											
__/__											
__/__											
__/__											
__/__											
__/__											
__/__											
		Totals this page:									I certify that the entries in this log are true.
		Totals brought forward:									
		Totals to date:									

21

Date	Aircraft	Location		Flight Duration				Function		Ld	Remarks and Endorsements
Yr___	Make/Model	Takeoff	Landing	Fixed-wing	Rotary	LTA	Total	OIC/Solo	Student/ Instructor	No.	
Dd/ Mm	Registration	Time	Time								
__/__											
__/__											
__/__											
__/__											
__/__											
__/__											
__/__											
__/__											
Totals this page:											*I certify that the entries in this log are true.*
Totals brought forward:											
Totals to date:											

DATE	AIRCRAFT	LOCATION		FLIGHT DURATION				FUNCTION		Ld	REMARKS AND ENDORSEMENTS
Yr	Make/Model	Takeoff	Landing	Fixed-wing	Rotary	LTA	Total	OIC/Solo	Student/ Instructor	No.	
Dd/ Mm	Registration	Time	Time								
__/__											
__/__											
__/__											
__/__											
__/__											
__/__											
__/__											
__/__											
		Totals this page:									I certify that the entries in this log are true.
		Totals brought forward:									
		Totals to date:									

23

DATE	AIRCRAFT	LOCATION		FLIGHT DURATION				FUNCTION		LD	REMARKS AND ENDORSEMENTS
Yr___	Make/Model	Takeoff	Landing	Fixed-wing	Rotary	LTA	Total	OIC/Solo	Student/ Instructor	No.	
Dd/ Mm	Registration	Time	Time								
__/__											
__/__											
__/__											
__/__											
__/__											
__/__											
__/__											
__/__											
		Totals this page:									*I certify that the entries in this log are true.*
		Totals brought forward:									
		Totals to date:									

DATE	AIRCRAFT	LOCATION		FLIGHT DURATION				FUNCTION		LD	REMARKS AND ENDORSEMENTS
Yr	Make/Model	Takeoff	Landing	Fixed-wing	Rotary	LTA	Total	OIC/Solo	Student/ Instructor	No.	
Dd/ Mm	Registration	Time	Time								
__/__											
__/__											
__/__											
__/__											
__/__											
__/__											
__/__											
__/__											
	Totals this page:										I certify that the entries in this log are true.
	Totals brought forward:										
	Totals to date:										

25

DATE	AIRCRAFT	LOCATION		FLIGHT DURATION				FUNCTION		LD	REMARKS AND ENDORSEMENTS
Yr	Make/Model	Takeoff	Landing	Fixed-wing	Rotary	LTA	Total	OIC/Solo	Student/ Instructor	No.	
Dd/ Mm	Registration	Time	Time								
__/__											
__/__											
__/__											
__/__											
__/__											
__/__											
__/__											
__/__											
	Totals this page:										I certify that the entries in this log are true.
	Totals brought forward:										
	Totals to date:										

DATE	AIRCRAFT	LOCATION		FLIGHT DURATION				FUNCTION		LD	REMARKS AND ENDORSEMENTS
Yr	Make/Model	Takeoff	Landing	Fixed-wing	Rotary	LTA	Total	OIC/Solo	Student/Instructor	No.	
Dd/ Mm	Registration	Time	Time								
__/__											
__/__											
__/__											
__/__											
__/__											
__/__											
__/__											
__/__											
		Totals this page:									*I certify that the entries in this log are true.*
		Totals brought forward:									
		Totals to date:									

Date	Aircraft	Location		Flight Duration				Function		Ld	Remarks and Endorsements
Yr	Make/Model	Takeoff	Landing	Fixed-wing	Rotary	LTA	Total	OIC/Solo	Student/Instructor	No.	
Dd/ Mm	Registration	Time	Time								
__/__											
__/__											
__/__											
__/__											
__/__											
__/__											
__/__											
__/__											
Totals this page:											*I certify that the entries in this log are true.*
Totals brought forward:											
Totals to date:											

28

DATE	AIRCRAFT	LOCATION		FLIGHT DURATION				FUNCTION		LD	REMARKS AND ENDORSEMENTS
Yr	Make/Model	Takeoff	Landing	Fixed-wing	Rotary	LTA	Total	OIC/Solo	Student/Instructor	No.	
Dd/ Mm	Registration	Time	Time								
__/__											
__/__											
__/__											
__/__											
__/__											
__/__											
__/__											
__/__											
		Totals this page:									I certify that the entries in this log are true.
		Totals brought forward:									
		Totals to date:									

29

Date	Aircraft	Location		Flight Duration				Function		LD	Remarks and Endorsements
Yr___	Make/Model	Takeoff	Landing	Fixed-wing	Rotary	LTA	Total	OIC/Solo	Student/Instructor	No.	
Dd/ Mm	Registration	Time	Time								
__/__											
__/__											
__/__											
__/__											
__/__											
__/__											
__/__											
__/__											
	Totals this page:										I certify that the entries in this log are true.
	Totals brought forward:										
	Totals to date:										

DATE	AIRCRAFT	LOCATION		FLIGHT DURATION				FUNCTION		LD	REMARKS AND ENDORSEMENTS
Yr	Make/Model	Takeoff	Landing	Fixed-wing	Rotary	LTA	Total	OIC/Solo	Student/Instructor	No.	
Dd/ Mm	Registration	Time	Time								
__/__											
__/__											
__/__											
__/__											
__/__											
__/__											
__/__											
__/__											
Totals this page:											I certify that the entries in this log are true.
Totals brought forward:											
Totals to date:											

Date	Aircraft	Location		Flight Duration				Function		Ld	Remarks and Endorsements
Yr___	Make/Model	Takeoff	Landing	Fixed-wing	Rotary	LTA	Total	OIC/Solo	Student/Instructor	No.	
Dd/ Mm	Registration	Time	Time								
__/__											
__/__											
__/__											
__/__											
__/__											
__/__											
__/__											
__/__											
Totals this page:											*I certify that the entries in this log are true.*
Totals brought forward:											
Totals to date:											

DATE	AIRCRAFT	LOCATION		FLIGHT DURATION				FUNCTION		LD	REMARKS AND ENDORSEMENTS
Yr	Make/Model	Takeoff	Landing	Fixed-wing	Rotary	LTA	Total	OIC/Solo	Student/Instructor	No.	
Dd/ Mm	Registration	Time	Time								
__/__											
__/__											
__/__											
__/__											
__/__											
__/__											
__/__											
__/__											
Totals this page:											I certify that the entries in this log are true.
Totals brought forward:											
Totals to date:											

33

DATE	AIRCRAFT	LOCATION		FLIGHT DURATION				FUNCTION		LD	REMARKS AND ENDORSEMENTS
Yr	Make/Model	Takeoff	Landing	Fixed-wing	Rotary	LTA	Total	OIC/Solo	Student/ Instructor	No.	
Dd/ Mm	Registration	Time	Time								
__/__											
__/__											
__/__											
__/__											
__/__											
__/__											
__/__											
__/__											
		Totals this page:									*I certify that the entries in this log are true.*
		Totals brought forward:									
		Totals to date:									

DATE	AIRCRAFT	LOCATION		FLIGHT DURATION				FUNCTION		LD	REMARKS AND ENDORSEMENTS
Yr	Make/Model	Takeoff	Landing	Fixed-wing	Rotary	LTA	Total	OIC/Solo	Student/ Instructor	No.	
Dd/ Mm	Registration	Time	Time								
__/__											
__/__											
__/__											
__/__											
__/__											
__/__											
__/__											
__/__											
__/__											
	Totals this page:										I certify that the entries in this log are true.
	Totals brought forward:										
	Totals to date:										

35

DATE	AIRCRAFT	LOCATION		FLIGHT DURATION				FUNCTION		LD	REMARKS AND ENDORSEMENTS
Yr	Make/Model	Takeoff	Landing	Fixed-wing	Rotary	LTA	Total	OIC/Solo	Student/Instructor	No.	
Dd/ Mm	Registration	Time	Time								
___/___											
___/___											
___/___											
___/___											
___/___											
___/___											
___/___											
___/___											
		Totals this page:									*I certify that the entries in this log are true.*
		Totals brought forward:									
		Totals to date:									

Date	Aircraft	Location		Flight Duration				Function		Ld	Remarks and Endorsements
Yr	Make/Model	Takeoff	Landing	Fixed-wing	Rotary	LTA	Total	OIC/Solo	Student/ Instructor	No.	
Dd/ Mm	Registration	Time	Time								
__/__											
__/__											
__/__											
__/__											
__/__											
__/__											
__/__											
__/__											
	Totals this page:										I certify that the entries in this log are true.
	Totals brought forward:										
	Totals to date:										

DATE	AIRCRAFT	LOCATION		FLIGHT DURATION				FUNCTION		LD	REMARKS AND ENDORSEMENTS
Yr	Make/Model	Takeoff	Landing	Fixed-wing	Rotary	LTA	Total	OIC/Solo	Student/ Instructor	No.	
Dd/ Mm	Registration	Time	Time								
__/__											
__/__											
__/__											
__/__											
__/__											
__/__											
__/__											
__/__											
Totals this page:											I certify that the entries in this log are true.
Totals brought forward:											
Totals to date:											

Date	Aircraft	Location		Flight Duration				Function		Ld	Remarks and Endorsements
Yr	Make/Model	Takeoff	Landing	Fixed-wing	Rotary	LTA	Total	OIC/Solo	Student/Instructor	No.	
Dd/ Mm	Registration	Time	Time								
__/__											
__/__											
__/__											
__/__											
__/__											
__/__											
__/__											
__/__											
	Totals this page:										*I certify that the entries in this log are true.*
	Totals brought forward:										
	Totals to date:										

39

Date	Aircraft	Location		Flight Duration				Function		Ld	Remarks and Endorsements
Yr___	Make/Model	Takeoff	Landing	Fixed-wing	Rotary	LTA	Total	OIC/Solo	Student/Instructor	No.	
Dd/ Mm	Registration	Time	Time								
__/__											
__/__											
__/__											
__/__											
__/__											
__/__											
__/__											
__/__											
Totals this page:											I certify that the entries in this log are true.
Totals brought forward:											
Totals to date:											

DATE	AIRCRAFT		LOCATION		FLIGHT DURATION				FUNCTION		LD	REMARKS AND ENDORSEMENTS
Yr	Make/Model		Takeoff	Landing	Fixed-wing	Rotary	LTA	Total	OIC/Solo	Student/ Instructor	No.	
Dd/ Mm	Registration		Time	Time								
__/__												
__/__												
__/__												
__/__												
__/__												
__/__												
__/__												
__/__												
	Totals this page:											I certify that the entries in this log are true.
	Totals brought forward:											
	Totals to date:											

41

DATE	AIRCRAFT	LOCATION		FLIGHT DURATION				FUNCTION		LD	REMARKS AND ENDORSEMENTS
Yr	Make/Model	Takeoff	Landing	Fixed-wing	Rotary	LTA	Total	OIC/Solo	Student/ Instructor	No.	
Dd/ Mm	Registration	Time	Time								
__/__											
__/__											
__/__											
__/__											
__/__											
__/__											
__/__											
__/__											
Totals this page:											I certify that the entries in this log are true.
Totals brought forward:											
Totals to date:											

DATE	AIRCRAFT	LOCATION		FLIGHT DURATION				FUNCTION		LD	REMARKS AND ENDORSEMENTS
Yr	Make/Model	Takeoff	Landing	Fixed-wing	Rotary	LTA	Total	OIC/Solo	Student/Instructor	No.	
Dd/ Mm	Registration	Time	Time								
__/__											
__/__											
__/__											
__/__											
__/__											
__/__											
__/__											
__/__											
		Totals this page:									*I certify that the entries in this log are true.*
		Totals brought forward:									
		Totals to date:									

43

Date	Aircraft	Location		Flight Duration				Function		Ld	Remarks and Endorsements
Yr	Make/Model	Takeoff	Landing	Fixed-wing	Rotary	LTA	Total	OIC/Solo	Student/Instructor	No.	
Dd/ Mm	Registration	Time	Time								
__/__											
__/__											
__/__											
__/__											
__/__											
__/__											
__/__											
__/__											
Totals this page:											*I certify that the entries in this log are true.*
Totals brought forward:											
Totals to date:											

44

DATE	AIRCRAFT		LOCATION		FLIGHT DURATION				FUNCTION		LD	REMARKS AND ENDORSEMENTS
Yr	Make/Model		Takeoff	Landing	Fixed-wing	Rotary	LTA	Total	OIC/Solo	Student/ Instructor	No.	
Dd/ Mm	Registration		Time	Time								
__/__												
__/__												
__/__												
__/__												
__/__												
__/__												
__/__												
__/__												
Totals this page:												*I certify that the entries in this log are true.*
Totals brought forward:												
Totals to date:												

DATE	AIRCRAFT	LOCATION		FLIGHT DURATION				FUNCTION		LD	REMARKS AND ENDORSEMENTS
Yr	Make/Model	Takeoff	Landing	Fixed-wing	Rotary	LTA	Total	OIC/Solo	Student/ Instructor	No.	
Dd/ Mm	Registration	Time	Time								
__/__											
__/__											
__/__											
__/__											
__/__											
__/__											
__/__											
__/__											
Totals this page:											*I certify that the entries in this log are true.*
Totals brought forward:											
Totals to date:											

46

Date	Aircraft	Location		Flight Duration				Function		Ld	Remarks and Endorsements
Yr___	Make/Model	Takeoff	Landing	Fixed-wing	Rotary	LTA	Total	OIC/Solo	Student/ Instructor	No.	
Dd/ Mm	Registration	Time	Time								
__/__											
__/__											
__/__											
__/__											
__/__											
__/__											
__/__											
__/__											
	Totals this page:										I certify that the entries in this log are true.
	Totals brought forward:										
	Totals to date:										

47

DATE	AIRCRAFT	LOCATION		FLIGHT DURATION				FUNCTION		LD	REMARKS AND ENDORSEMENTS
Yr	Make/Model	Takeoff	Landing	Fixed-wing	Rotary	LTA	Total	OIC/Solo	Student/ Instructor	No.	
Dd/ Mm	Registration	Time	Time								
__/__											
__/__											
__/__											
__/__											
__/__											
__/__											
__/__											
__/__											
		Totals this page:									I certify that the entries in this log are true.
		Totals brought forward:									
		Totals to date:									

DATE	AIRCRAFT	LOCATION		FLIGHT DURATION				FUNCTION		LD	REMARKS AND ENDORSEMENTS
Yr	Make/Model	Takeoff	Landing	Fixed-wing	Rotary	LTA	Total	OIC/Solo	Student/Instructor	No.	
Dd/ Mm	Registration	Time	Time								
__/__											
__/__											
__/__											
__/__											
__/__											
__/__											
__/__											
__/__											
Totals this page:											I certify that the entries in this log are true.
Totals brought forward:											
Totals to date:											

49

DATE	AIRCRAFT	LOCATION		FLIGHT DURATION				FUNCTION		LD	REMARKS AND ENDORSEMENTS
Yr	Make/Model	Takeoff	Landing	Fixed-wing	Rotary	LTA	Total	OIC/Solo	Student/ Instructor	No.	
Dd/ Mm	Registration	Time	Time								
__ / __											
__ / __											
__ / __											
__ / __											
__ / __											
__ / __											
__ / __											
__ / __											
	Totals this page:										I certify that the entries in this log are true.
	Totals brought forward:										
	Totals to date:										

50

Date	Aircraft	Location		Flight Duration				Function		Ld	Remarks and Endorsements
Yr	Make/Model	Takeoff	Landing	Fixed-wing	Rotary	LTA	Total	OIC/Solo	Student/Instructor	No.	
Dd/ Mm	Registration	Time	Time								
__/__											
__/__											
__/__											
__/__											
__/__											
__/__											
__/__											
__/__											
Totals this page:											*I certify that the entries in this log are true.*
Totals brought forward:											
Totals to date:											

51

Date	Aircraft	Location		Flight Duration				Function		Ld	Remarks and Endorsements
Yr	Make/Model	Takeoff	Landing	Fixed-wing	Rotary	LTA	Total	OIC/Solo	Student/Instructor	No.	
Dd/ Mm	Registration	Time	Time								
__/__											
__/__											
__/__											
__/__											
__/__											
__/__											
__/__											
__/__											
Totals this page:											*I certify that the entries in this log are true.*
Totals brought forward:											
Totals to date:											

DATE	AIRCRAFT	LOCATION		FLIGHT DURATION				FUNCTION		LD	REMARKS AND ENDORSEMENTS
Yr	Make/Model	Takeoff	Landing	Fixed-wing	Rotary	LTA	Total	OIC/Solo	Student/Instructor	No.	
Dd/ Mm	Registration	Time	Time								
__/__											
__/__											
__/__											
__/__											
__/__											
__/__											
__/__											
__/__											
		Totals this page:									I certify that the entries in this log are true.
		Totals brought forward:									
		Totals to date:									

Date	Aircraft	Location		Flight Duration				Function		Ld	Remarks and Endorsements
Yr	Make/Model	Takeoff	Landing	Fixed-wing	Rotary	LTA	Total	OIC/Solo	Student/ Instructor	No.	
Dd/ Mm	Registration	Time	Time								
__/__											
__/__											
__/__											
__/__											
__/__											
__/__											
__/__											
__/__											
Totals this page:											*I certify that the entries in this log are true.*
Totals brought forward:											
Totals to date:											

Date	Aircraft		Location		Flight Duration				Function		LD	Remarks and Endorsements
Yr	Make/Model		Takeoff	Landing	Fixed-wing	Rotary	LTA	Total	OIC/Solo	Student/Instructor	No.	
Dd/ Mm	Registration		Time	Time								
__/__												
__/__												
__/__												
__/__												
__/__												
__/__												
__/__												
__/__												
Totals this page:												I certify that the entries in this log are true.
Totals brought forward:												
Totals to date:												

55

DATE	AIRCRAFT	LOCATION		FLIGHT DURATION				FUNCTION		LD	REMARKS AND ENDORSEMENTS
Yr	Make/Model	Takeoff	Landing	Fixed-wing	Rotary	LTA	Total	OIC/Solo	Student/Instructor	No.	
Dd/ Mm	Registration	Time	Time								
__/__											
__/__											
__/__											
__/__											
__/__											
__/__											
__/__											
__/__											
Totals this page:											I certify that the entries in this log are true.
Totals brought forward:											
Totals to date:											

Date	Aircraft	Location		Flight Duration				Function		Ld	Remarks and Endorsements
Yr	Make/Model	Takeoff	Landing	Fixed-wing	Rotary	LTA	Total	OIC/Solo	Student/Instructor	No.	
Dd/ Mm	Registration	Time	Time								
__/__											
__/__											
__/__											
__/__											
__/__											
__/__											
__/__											
__/__											
Totals this page:											I certify that the entries in this log are true.
Totals brought forward:											
Totals to date:											

DATE	AIRCRAFT	LOCATION		FLIGHT DURATION				FUNCTION		LD	REMARKS AND ENDORSEMENTS
Yr	Make/Model	Takeoff	Landing	Fixed-wing	Rotary	LTA	Total	OIC/Solo	Student/ Instructor	No.	
Dd/ Mm	Registration	Time	Time								
__ / __											
__ / __											
__ / __											
__ / __											
__ / __											
__ / __											
__ / __											
__ / __											
Totals this page:											*I certify that the entries in this log are true.*
Totals brought forward:											
Totals to date:											

58

DATE	AIRCRAFT	LOCATION		FLIGHT DURATION				FUNCTION		LD	REMARKS AND ENDORSEMENTS
Yr	Make/Model	Takeoff	Landing	Fixed-wing	Rotary	LTA	Total	OIC/Solo	Student/ Instructor	No.	
Dd/ Mm	Registration	Time	Time								
__/__											
__/__											
__/__											
__/__											
__/__											
__/__											
__/__											
__/__											
	Totals this page:										*I certify that the entries in this log are true.*
	Totals brought forward:										
	Totals to date:										

59

Date	Aircraft	Location		Flight Duration				Function		Ld	Remarks and Endorsements
Yr	Make/Model	Takeoff	Landing	Fixed-wing	Rotary	LTA	Total	OIC/Solo	Student/Instructor	No.	
Dd/ Mm	Registration	Time	Time								
__/__											
__/__											
__/__											
__/__											
__/__											
__/__											
__/__											
__/__											
Totals this page:											*I certify that the entries in this log are true.*
Totals brought forward:											
Totals to date:											

DATE	AIRCRAFT	LOCATION		FLIGHT DURATION				FUNCTION		LD	REMARKS AND ENDORSEMENTS
Yr	Make/Model	Takeoff	Landing	Fixed-wing	Rotary	LTA	Total	OIC/Solo	Student/ Instructor	No.	
Dd/ Mm	Registration	Time	Time								
__/__											
__/__											
__/__											
__/__											
__/__											
__/__											
__/__											
__/__											
		Totals this page:									I certify that the entries in this log are true.
		Totals brought forward:									
		Totals to date:									

DATE	AIRCRAFT	LOCATION		FLIGHT DURATION				FUNCTION		LD	REMARKS AND ENDORSEMENTS
Yr___	Make/Model	Takeoff	Landing	Fixed-wing	Rotary	LTA	Total	OIC/Solo	Student/ Instructor	No.	
Dd/ Mm	Registration	Time	Time								
__/__											
__/__											
__/__											
__/__											
__/__											
__/__											
__/__											
__/__											
		Totals this page:									I certify that the entries in this log are true.
		Totals brought forward:									
		Totals to date:									

Date	Aircraft	Location		Flight Duration				Function		Ld	Remarks and Endorsements
Yr	Make/Model	Takeoff	Landing	Fixed-wing	Rotary	LTA	Total	OIC/Solo	Student/Instructor	No.	
Dd/ Mm	Registration	Time	Time								
__/__											
__/__											
__/__											
__/__											
__/__											
__/__											
__/__											
__/__											
		Totals this page:									*I certify that the entries in this log are true.*
		Totals brought forward:									
		Totals to date:									

63

DATE	AIRCRAFT	LOCATION		FLIGHT DURATION				FUNCTION		LD	REMARKS AND ENDORSEMENTS
Yr	Make/Model	Takeoff	Landing	Fixed-wing	Rotary	LTA	Total	OIC/Solo	Student/ Instructor	No.	
Dd/ Mm	Registration	Time	Time								
___/___											
___/___											
___/___											
___/___											
___/___											
___/___											
___/___											
___/___											
Totals this page:											*I certify that the entries in this log are true.*
Totals brought forward:											
Totals to date:											

64

DATE	AIRCRAFT	LOCATION		FLIGHT DURATION				FUNCTION		LD	REMARKS AND ENDORSEMENTS
Yr	Make/Model	Takeoff	Landing	Fixed-wing	Rotary	LTA	Total	OIC/Solo	Student/Instructor	No.	
Dd/ Mm	Registration	Time	Time								
__/__											
__/__											
__/__											
__/__											
__/__											
__/__											
__/__											
__/__											
Totals this page:											I certify that the entries in this log are true.
Totals brought forward:											
Totals to date:											

65

Date	Aircraft	Location		Flight Duration				Function		Ld	Remarks and Endorsements
Yr	Make/Model	Takeoff	Landing	Fixed-wing	Rotary	LTA	Total	OIC/Solo	Student/ Instructor	No.	
Dd/ Mm	Registration	Time	Time								
__/__											
__/__											
__/__											
__/__											
__/__											
__/__											
__/__											
__/__											
Totals this page:											I certify that the entries in this log are true.
Totals brought forward:											
Totals to date:											

DATE	AIRCRAFT	LOCATION		FLIGHT DURATION				FUNCTION		LD	REMARKS AND ENDORSEMENTS
Yr	Make/Model	Takeoff	Landing	Fixed-wing	Rotary	LTA	Total	OIC/Solo	Student/Instructor	No.	
Dd/ Mm	Registration	Time	Time								
__/__											
__/__											
__/__											
__/__											
__/__											
__/__											
__/__											
__/__											
		Totals this page:									I certify that the entries in this log are true.
		Totals brought forward:									
		Totals to date:									

DATE	AIRCRAFT	LOCATION		FLIGHT DURATION				FUNCTION		LD	REMARKS AND ENDORSEMENTS
Yr	Make/Model	Takeoff	Landing	Fixed-wing	Rotary	LTA	Total	OIC/Solo	Student/ Instructor	No.	
Dd/ Mm	Registration	Time	Time								
__/__											
__/__											
__/__											
__/__											
__/__											
__/__											
__/__											
__/__											
Totals this page:											*I certify that the entries in this log are true.*
Totals brought forward:											
Totals to date:											

69

DATE	AIRCRAFT	LOCATION		FLIGHT DURATION				FUNCTION		LD	REMARKS AND ENDORSEMENTS
Yr	Make/Model	Takeoff	Landing	Fixed-wing	Rotary	LTA	Total	OIC/Solo	Student/ Instructor	No.	
Dd/ Mm	Registration	Time	Time								
__/__											
__/__											
__/__											
__/__											
__/__											
__/__											
__/__											
__/__											
		Totals this page:									*I certify that the entries in this log are true.*
		Totals brought forward:									
		Totals to date:									

Date	Aircraft	Location		Flight Duration				Function		Ld	Remarks and Endorsements
Yr___	Make/Model	Takeoff	Landing	Fixed-wing	Rotary	LTA	Total	OIC/Solo	Student/ Instructor	No.	
Dd/ Mm	Registration	Time	Time								
__/__											
__/__											
__/__											
__/__											
__/__											
__/__											
__/__											
__/__											
Totals this page:											*I certify that the entries in this log are true.*
Totals brought forward:											
Totals to date:											

70

DATE	AIRCRAFT	LOCATION		FLIGHT DURATION				FUNCTION		LD	REMARKS AND ENDORSEMENTS
Yr	Make/Model	Takeoff	Landing	Fixed-wing	Rotary	LTA	Total	OIC/Solo	Student/ Instructor	No.	
Dd/ Mm	Registration	Time	Time								
__/__											
__/__											
__/__											
__/__											
__/__											
__/__											
__/__											
__/__											
		Totals this page:									I certify that the entries in this log are true.
		Totals brought forward:									
		Totals to date:									

71

DATE	AIRCRAFT	LOCATION		FLIGHT DURATION				FUNCTION		LD	REMARKS AND ENDORSEMENTS
Yr	Make/Model	Takeoff	Landing	Fixed-wing	Rotary	LTA	Total	OIC/Solo	Student/ Instructor	No.	
Dd/ Mm	Registration	Time	Time								
__/__											
__/__											
__/__											
__/__											
__/__											
__/__											
__/__											
__/__											
		Totals this page:									*I certify that the entries in this log are true.*
		Totals brought forward:									
		Totals to date:									

DATE	AIRCRAFT	LOCATION		FLIGHT DURATION				FUNCTION		LD	REMARKS AND ENDORSEMENTS
Yr	Make/Model	Takeoff	Landing	Fixed-wing	Rotary	LTA	Total	OIC/Solo	Student/ Instructor	No.	
Dd/ Mm	Registration	Time	Time								
__/__											
__/__											
__/__											
__/__											
__/__											
__/__											
__/__											
__/__											
Totals this page:											I certify that the entries in this log are true.
Totals brought forward:											
Totals to date:											

73

DATE	AIRCRAFT	LOCATION		FLIGHT DURATION				FUNCTION		LD	REMARKS AND ENDORSEMENTS
Yr	Make/Model	Takeoff	Landing	Fixed-wing	Rotary	LTA	Total	OIC/Solo	Student/Instructor	No.	
Dd/ Mm	Registration	Time	Time								
__/__											
__/__											
__/__											
__/__											
__/__											
__/__											
__/__											
__/__											
Totals this page:											I certify that the entries in this log are true.
Totals brought forward:											
Totals to date:											

DATE	AIRCRAFT	LOCATION		FLIGHT DURATION				FUNCTION		LD	REMARKS AND ENDORSEMENTS
Yr	Make/Model	Takeoff	Landing	Fixed-wing	Rotary	LTA	Total	OIC/Solo	Student/Instructor	No.	
Dd/ Mm	Registration	Time	Time								
__ / __											
__ / __											
__ / __											
__ / __											
__ / __											
__ / __											
__ / __											
__ / __											
	Totals this page:										*I certify that the entries in this log are true.*
	Totals brought forward:										
	Totals to date:										

75

Date	Aircraft	Location		Flight Duration				Function		Ld	Remarks and Endorsements
Yr	Make/Model	Takeoff	Landing	Fixed-wing	Rotary	LTA	Total	OIC/Solo	Student/Instructor	No.	
Dd/ Mm	Registration	Time	Time								
__/__											
__/__											
__/__											
__/__											
__/__											
__/__											
__/__											
__/__											
Totals this page:											*I certify that the entries in this log are true.*
Totals brought forward:											
Totals to date:											

DATE	AIRCRAFT	LOCATION		FLIGHT DURATION				FUNCTION		LD	REMARKS AND ENDORSEMENTS
Yr	Make/Model	Takeoff	Landing	Fixed-wing	Rotary	LTA	Total	OIC/Solo	Student/ Instructor	No.	
Dd/ Mm	Registration	Time	Time								
__/__											
__/__											
__/__											
__/__											
__/__											
__/__											
__/__											
__/__											
	Totals this page:										*I certify that the entries in this log are true.*
	Totals brought forward:										
	Totals to date:										

DATE	AIRCRAFT	LOCATION		FLIGHT DURATION				FUNCTION		LD	REMARKS AND ENDORSEMENTS
Yr	Make/Model	Takeoff	Landing	Fixed-wing	Rotary	LTA	Total	OIC/Solo	Student/ Instructor	No.	
Dd/ Mm	Registration	Time	Time								
__/__											
__/__											
__/__											
__/__											
__/__											
__/__											
__/__											
__/__											
Totals this page:											I certify that the entries in this log are true.
Totals brought forward:											
Totals to date:											

DATE	AIRCRAFT	LOCATION		FLIGHT DURATION				FUNCTION		LD	REMARKS AND ENDORSEMENTS
Yr	Make/Model	Takeoff	Landing	Fixed-wing	Rotary	LTA	Total	OIC/Solo	Student/ Instructor	No.	
Dd/ Mm	Registration	Time	Time								
__/__											
__/__											
__/__											
__/__											
__/__											
__/__											
__/__											
Totals this page:											I certify that the entries in this log are true.
Totals brought forward:											
Totals to date:											

DATE	AIRCRAFT	LOCATION		FLIGHT DURATION				FUNCTION		LD	REMARKS AND ENDORSEMENTS
Yr	Make/Model	Takeoff	Landing	Fixed-wing	Rotary	LTA	Total	OIC/Solo	Student/Instructor	No.	
Dd/ Mm	Registration	Time	Time								
__/__											
__/__											
__/__											
__/__											
__/__											
__/__											
__/__											
__/__											
Totals this page:											I certify that the entries in this log are true.
Totals brought forward:											
Totals to date:											

DATE	AIRCRAFT		LOCATION		FLIGHT DURATION				FUNCTION		LD	REMARKS AND ENDORSEMENTS
Yr	Make/Model		Takeoff	Landing	Fixed-wing	Rotary	LTA	Total	OIC/Solo	Student/ Instructor	No.	
Dd/ Mm	Registration		Time	Time								
__/__												
__/__												
__/__												
__/__												
__/__												
__/__												
__/__												
__/__												
Totals this page:												*I certify that the entries in this log are true.*
Totals brought forward:												
Totals to date:												

81

DATE	AIRCRAFT	LOCATION		FLIGHT DURATION				FUNCTION		LD	REMARKS AND ENDORSEMENTS
Yr	Make/Model	Takeoff	Landing	Fixed-wing	Rotary	LTA	Total	OIC/Solo	Student/ Instructor	No.	
Dd/ Mm	Registration	Time	Time								
__/__											
__/__											
__/__											
__/__											
__/__											
__/__											
__/__											
__/__											
Totals this page:											*I certify that the entries in this log are true.*
Totals brought forward:											
Totals to date:											

82

DATE	AIRCRAFT	LOCATION		FLIGHT DURATION				FUNCTION		LD	REMARKS AND ENDORSEMENTS
Yr	Make/Model	Takeoff	Landing	Fixed-wing	Rotary	LTA	Total	OIC/Solo	Student/Instructor	No.	
Dd/ Mm	Registration	Time	Time								
__/__											
__/__											
__/__											
__/__											
__/__											
__/__											
__/__											
__/__											
Totals this page:											I certify that the entries in this log are true.
Totals brought forward:											
Totals to date:											

DATE	AIRCRAFT	LOCATION		FLIGHT DURATION				FUNCTION		LD	REMARKS AND ENDORSEMENTS
Yr	Make/Model	Takeoff	Landing	Fixed-wing	Rotary	LTA	Total	OIC/Solo	Student/ Instructor	No.	
Dd/ Mm	Registration	Time	Time								
___/___											
___/___											
___/___											
___/___											
___/___											
___/___											
___/___											
___/___											
Totals this page:											*I certify that the entries in this log are true.*
Totals brought forward:											
Totals to date:											

DATE	AIRCRAFT	LOCATION		FLIGHT DURATION				FUNCTION		LD	REMARKS AND ENDORSEMENTS
Yr	Make/Model	Takeoff	Landing	Fixed-wing	Rotary	LTA	Total	OIC/Solo	Student/ Instructor	No.	
Dd/ Mm	Registration	Time	Time								
__/__											
__/__											
__/__											
__/__											
__/__											
__/__											
__/__											
__/__											
		Totals this page:									I certify that the entries in this log are true.
		Totals brought forward:									
		Totals to date:									

DATE	AIRCRAFT	LOCATION		FLIGHT DURATION				FUNCTION		LD	REMARKS AND ENDORSEMENTS
Yr	Make/Model	Takeoff	Landing	Fixed-wing	Rotary	LTA	Total	OIC/Solo	Student/ Instructor	No.	
Dd/ Mm	Registration	Time	Time								
__/__											
__/__											
__/__											
__/__											
__/__											
__/__											
__/__											
__/__											
Totals this page:											I certify that the entries in this log are true.
Totals brought forward:											
Totals to date:											

Date	Aircraft	Location		Flight Duration				Function		Ld	Remarks and Endorsements
Yr	Make/Model	Takeoff	Landing	Fixed-wing	Rotary	LTA	Total	OIC/Solo	Student/Instructor	No.	
Dd/ Mm	Registration	Time	Time								
__/__											
__/__											
__/__											
__/__											
__/__											
__/__											
__/__											
__/__											
	Totals this page:										*I certify that the entries in this log are true.*
	Totals brought forward:										
	Totals to date:										

87

Date	Aircraft	Location		Flight Duration				Function		Ld	Remarks and Endorsements
Yr	Make/Model	Takeoff	Landing	Fixed-wing	Rotary	LTA	Total	OIC/Solo	Student/Instructor	No.	
Dd/ Mm	Registration	Time	Time								
__/__											
__/__											
__/__											
__/__											
__/__											
__/__											
__/__											
__/__											
	Totals this page:										*I certify that the entries in this log are true.*
	Totals brought forward:										
	Totals to date:										

DATE	AIRCRAFT	LOCATION		FLIGHT DURATION				FUNCTION		LD	REMARKS AND ENDORSEMENTS
Yr	Make/Model	Takeoff	Landing	Fixed-wing	Rotary	LTA	Total	OIC/Solo	Student/ Instructor	No.	
Dd/ Mm	Registration	Time	Time								
__/__											
__/__											
__/__											
__/__											
__/__											
__/__											
__/__											
__/__											
		Totals this page:									I certify that the entries in this log are true.
		Totals brought forward:									
		Totals to date:									

Date	Aircraft	Location		Flight Duration				Function		Ld	Remarks and Endorsements
Yr	Make/Model	Takeoff	Landing	Fixed-wing	Rotary	LTA	Total	OIC/Solo	Student/ Instructor	No.	
Dd/ Mm	Registration	Time	Time								
__/__											
__/__											
__/__											
__/__											
__/__											
__/__											
__/__											
__/__											
		Totals this page:									*I certify that the entries in this log are true.*
		Totals brought forward:									
		Totals to date:									

DATE	AIRCRAFT		LOCATION		FLIGHT DURATION				FUNCTION		LD	REMARKS AND ENDORSEMENTS
Yr	Make/Model		Takeoff	Landing	Fixed-wing	Rotary	LTA	Total	OIC/Solo	Student/ Instructor	No.	
Dd/ Mm	Registration		Time	Time								
__/__												
__/__												
__/__												
__/__												
__/__												
__/__												
__/__												
__/__												
			Totals this page:									*I certify that the entries in this log are true.*
			Totals brought forward:									
			Totals to date:									

91

Date	Aircraft	Location		Flight Duration				Function		Ld	Remarks and Endorsements
Yr ___	Make/Model	Takeoff	Landing	Fixed-wing	Rotary	LTA	Total	OIC/Solo	Student/ Instructor	No.	
Dd/ Mm	Registration	Time	Time								
__/__											
__/__											
__/__											
__/__											
__/__											
__/__											
__/__											
__/__											
Totals this page:											*I certify that the entries in this log are true.*
Totals brought forward:											
Totals to date:											

DATE	AIRCRAFT	LOCATION		FLIGHT DURATION				FUNCTION		LD	REMARKS AND ENDORSEMENTS
Yr	Make/Model	Takeoff	Landing	Fixed-wing	Rotary	LTA	Total	OIC/Solo	Student/Instructor	No.	
Dd/ Mm	Registration	Time	Time								
__/__											
__/__											
__/__											
__/__											
__/__											
__/__											
__/__											
__/__											
Totals this page:											I certify that the entries in this log are true.
Totals brought forward:											
Totals to date:											

Date	Aircraft	Location		Flight Duration				Function		LD	Remarks and Endorsements
Yr	Make/Model	Takeoff	Landing	Fixed-wing	Rotary	LTA	Total	OIC/Solo	Student/Instructor	No.	
Dd/ Mm	Registration	Time	Time								
__/__											
__/__											
__/__											
__/__											
__/__											
__/__											
__/__											
__/__											
Totals this page:											I certify that the entries in this log are true.
Totals brought forward:											
Totals to date:											

DATE	AIRCRAFT	LOCATION		FLIGHT DURATION				FUNCTION		LD	REMARKS AND ENDORSEMENTS
Yr	Make/Model	Takeoff	Landing	Fixed-wing	Rotary	LTA	Total	OIC/Solo	Student/Instructor	No.	
Dd/ Mm	Registration	Time	Time								
__/__											
__/__											
__/__											
__/__											
__/__											
__/__											
__/__											
__/__											
		Totals this page:									I certify that the entries in this log are true.
		Totals brought forward:									
		Totals to date:									

Date	Aircraft		Location		Flight Duration				Function		Ld	Remarks and Endorsements
Yr	Make/Model		Takeoff	Landing	Fixed-wing	Rotary	LTA	Total	OIC/Solo	Student/ Instructor	No.	
Dd/ Mm	Registration		Time	Time								
__/__												
__/__												
__/__												
__/__												
__/__												
__/__												
__/__												
__/__												
Totals this page:												I certify that the entries in this log are true.
Totals brought forward:												
Totals to date:												

DATE	AIRCRAFT	LOCATION		FLIGHT DURATION				FUNCTION		LD	REMARKS AND ENDORSEMENTS
Yr	Make/Model	Takeoff	Landing	Fixed-wing	Rotary	LTA	Total	OIC/Solo	Student/ Instructor	No.	
Dd/ Mm	Registration	Time	Time								
__/__											
__/__											
__/__											
__/__											
__/__											
__/__											
__/__											
__/__											
Totals this page:											I certify that the entries in this log are true.
Totals brought forward:											
Totals to date:											

DATE	AIRCRAFT		LOCATION		FLIGHT DURATION				FUNCTION		LD	REMARKS AND ENDORSEMENTS
Yr___	Make/Model		Takeoff	Landing	Fixed-wing	Rotary	LTA	Total	OIC/Solo	Student/ Instructor	No.	
Dd/ Mm	Registration		Time	Time								
___/___												
___/___												
___/___												
___/___												
___/___												
___/___												
___/___												
___/___												
Totals this page:												I certify that the entries in this log are true.
Totals brought forward:												
Totals to date:												

Date	Aircraft		Location		Flight Duration				Function		Ld	Remarks and Endorsements
Yr	Make/Model		Takeoff	Landing	Fixed-wing	Rotary	LTA	Total	OIC/Solo	Student/Instructor	No.	
Dd/ Mm	Registration		Time	Time								
__/__												
__/__												
__/__												
__/__												
__/__												
__/__												
__/__												
__/__												
			Totals this page:									*I certify that the entries in this log are true.*
			Totals brought forward:									
			Totals to date:									

Date	Aircraft	Location		Flight Duration				Function		Ld	Remarks and Endorsements
Yr	Make/Model	Takeoff	Landing	Fixed-wing	Rotary	LTA	Total	OIC/Solo	Student/Instructor	No.	
Dd/ Mm	Registration	Time	Time								
__/__											
__/__											
__/__											
__/__											
__/__											
__/__											
__/__											
__/__											
		Totals this page:									*I certify that the entries in this log are true.*
		Totals brought forward:									
		Totals to date:									

DATE	AIRCRAFT	LOCATION		FLIGHT DURATION				FUNCTION		LD	REMARKS AND ENDORSEMENTS
Yr	Make/Model	Takeoff	Landing	Fixed-wing	Rotary	LTA	Total	OIC/Solo	Student/ Instructor	No.	
Dd/ Mm	Registration	Time	Time								
__/__											
__/__											
__/__											
__/__											
__/__											
__/__											
__/__											
__/__											
		Totals this page:									I certify that the entries in this log are true.
		Totals brought forward:									
		Totals to date:									

DATE	AIRCRAFT	LOCATION		FLIGHT DURATION				FUNCTION		LD	REMARKS AND ENDORSEMENTS
Yr	Make/Model	Takeoff	Landing	Fixed-wing	Rotary	LTA	Total	OIC/Solo	Student/ Instructor	No.	
Dd/ Mm	Registration	Time	Time								
__/__											
__/__											
__/__											
__/__											
__/__											
__/__											
__/__											
Totals this page:											I certify that the entries in this log are true.
Totals brought forward:											
Totals to date:											

Date	Aircraft	Location		Flight Duration				Function		Ld	Remarks and Endorsements
Yr___	Make/Model	Takeoff	Landing	Fixed-wing	Rotary	LTA	Total	OIC/Solo	Student/Instructor	No.	
Dd/ Mm	Registration	Time	Time								
__/__											
__/__											
__/__											
__/__											
__/__											
__/__											
__/__											
__/__											
	Totals this page:										*I certify that the entries in this log are true.*
	Totals brought forward:										
	Totals to date:										

Date	Aircraft		Location		Flight Duration				Function		LD	Remarks and Endorsements
Yr___	Make/Model		Takeoff	Landing	Fixed-wing	Rotary	LTA	Total	OIC/Solo	Student/ Instructor	No.	
Dd/ Mm	Registration		Time	Time								
__/__												
__/__												
__/__												
__/__												
__/__												
__/__												
__/__												
__/__												
Totals this page:												*I certify that the entries in this log are true.*
Totals brought forward:												
Totals to date:												

Date	Aircraft	Location		Flight Duration				Function		Ld	Remarks and Endorsements
Yr	Make/Model	Takeoff	Landing	Fixed-wing	Rotary	LTA	Total	OIC/Solo	Student/Instructor	No.	
Dd/ Mm	Registration	Time	Time								
__/__											
__/__											
__/__											
__/__											
__/__											
__/__											
__/__											
__/__											
	Totals this page:										*I certify that the entries in this log are true.*
	Totals brought forward:										
	Totals to date:										

105

Date	Aircraft		Location			Flight Duration				Function			Ld	Remarks and Endorsements
Yr	Make/Model		Takeoff	Landing		Fixed-wing	Rotary	LTA	Total	OIC/Solo	Student/Instructor		No.	
Dd/ Mm	Registration		Time	Time										
__/__														
__/__														
__/__														
__/__														
__/__														
__/__														
__/__														
__/__														
Totals this page:														*I certify that the entries in this log are true.*
Totals brought forward:														
Totals to date:														

DATE	AIRCRAFT	LOCATION		FLIGHT DURATION				FUNCTION		LD	REMARKS AND ENDORSEMENTS
Yr	Make/Model	Takeoff	Landing	Fixed-wing	Rotary	LTA	Total	OIC/Solo	Student/ Instructor	No.	
Dd/ Mm	Registration	Time	Time								
__/__											
__/__											
__/__											
__/__											
__/__											
__/__											
__/__											
__/__											
Totals this page:											I certify that the entries in this log are true.
Totals brought forward:											
Totals to date:											

Date	Aircraft	Location		Flight Duration				Function		Ld	Remarks and Endorsements
Yr ___	Make/Model	Takeoff	Landing	Fixed-wing	Rotary	LTA	Total	OIC/Solo	Student/Instructor	No.	
Dd/ Mm	Registration	Time	Time								
__/__											
__/__											
__/__											
__/__											
__/__											
__/__											
__/__											
__/__											
Totals this page:											*I certify that the entries in this log are true.*
Totals brought forward:											
Totals to date:											

DATE	AIRCRAFT	LOCATION		FLIGHT DURATION				FUNCTION		LD	REMARKS AND ENDORSEMENTS
Yr	Make/Model	Takeoff	Landing	Fixed-wing	Rotary	LTA	Total	OIC/Solo	Student/ Instructor	No.	
Dd/ Mm	Registration	Time	Time								
__/__											
__/__											
__/__											
__/__											
__/__											
__/__											
__/__											
__/__											
Totals this page:											I certify that the entries in this log are true.
Totals brought forward:											
Totals to date:											

DATE	AIRCRAFT	LOCATION		FLIGHT DURATION				FUNCTION		LD	REMARKS AND ENDORSEMENTS
Yr	Make/Model	Takeoff	Landing	Fixed-wing	Rotary	LTA	Total	OIC/Solo	Student/ Instructor	No.	
Dd/ Mm	Registration	Time	Time								
__/__											
__/__											
__/__											
__/__											
__/__											
__/__											
__/__											
__/__											
Totals this page:											I certify that the entries in this log are true.
Totals brought forward:											
Totals to date:											

Date	Aircraft		Location		Flight Duration				Function		Ld	Remarks and Endorsements
Yr	Make/Model		Takeoff	Landing	Fixed-wing	Rotary	LTA	Total	OIC/Solo	Student/Instructor	No.	
Dd/ Mm	Registration		Time	Time								
__/__												
__/__												
__/__												
__/__												
__/__												
__/__												
__/__												
__/__												
Totals this page:												I certify that the entries in this log are true.
Totals brought forward:												
Totals to date:												

111

Date	Aircraft	Location		Flight Duration				Function		Ld	Remarks and Endorsements
Yr	Make/Model	Takeoff	Landing	Fixed-wing	Rotary	LTA	Total	OIC/Solo	Student/Instructor	No.	
Dd/ Mm	Registration	Time	Time								
__/__											
__/__											
__/__											
__/__											
__/__											
__/__											
__/__											
__/__											
			Totals this page:								I certify that the entries in this log are true.
			Totals brought forward:								
			Totals to date:								

DATE	AIRCRAFT	LOCATION		FLIGHT DURATION				FUNCTION		LD	REMARKS AND ENDORSEMENTS
Yr	Make/Model	Takeoff	Landing	Fixed-wing	Rotary	LTA	Total	OIC/Solo	Student/Instructor	No.	
Dd/ Mm	Registration	Time	Time								
__/__											
__/__											
__/__											
__/__											
__/__											
__/__											
__/__											
__/__											
Totals this page:											I certify that the entries in this log are true.
Totals brought forward:											
Totals to date:											

Date	Aircraft	Location		Flight Duration				Function		Ld	Remarks and Endorsements
Yr	Make/Model	Takeoff	Landing	Fixed-wing	Rotary	LTA	Total	OIC/Solo	Student/Instructor	No.	
Dd/ Mm	Registration	Time	Time								
__/__											
__/__											
__/__											
__/__											
__/__											
__/__											
__/__											
__/__											
Totals this page:											*I certify that the entries in this log are true.*
Totals brought forward:											
Totals to date:											

Date	Aircraft		Location		Flight Duration				Function		Ld	Remarks and Endorsements
Yr	Make/Model		Takeoff	Landing	Fixed-wing	Rotary	LTA	Total	OIC/Solo	Student/Instructor	No.	
Dd/ Mm	Registration		Time	Time								
__/__												
__/__												
__/__												
__/__												
__/__												
__/__												
__/__												
__/__												
			Totals this page:									I certify that the entries in this log are true.
			Totals brought forward:									
			Totals to date:									

115

DATE	AIRCRAFT	LOCATION		FLIGHT DURATION				FUNCTION		LD	REMARKS AND ENDORSEMENTS
Yr	Make/Model	Takeoff	Landing	Fixed-wing	Rotary	LTA	Total	OIC/Solo	Student/ Instructor	No.	
Dd/ Mm	Registration	Time	Time								
__/__											
__/__											
__/__											
__/__											
__/__											
__/__											
__/__											
__/__											
	Totals this page:										*I certify that the entries in this log are true.*
	Totals brought forward:										
	Totals to date:										

DATE	AIRCRAFT	LOCATION		FLIGHT DURATION				FUNCTION		LD	REMARKS AND ENDORSEMENTS
Yr	Make/Model	Takeoff	Landing	Fixed-wing	Rotary	LTA	Total	OIC/Solo	Student/Instructor	No.	
Dd/ Mm	Registration	Time	Time								
__/__											
__/__											
__/__											
__/__											
__/__											
__/__											
__/__											
__/__											
Totals this page:											I certify that the entries in this log are true.
Totals brought forward:											
Totals to date:											

Date	Aircraft	Location		Flight Duration				Function		LD	Remarks and Endorsements
Yr	Make/Model	Takeoff	Landing	Fixed-wing	Rotary	LTA	Total	OIC/Solo	Student/Instructor	No.	
Dd/ Mm	Registration	Time	Time								
__/__											
__/__											
__/__											
__/__											
__/__											
__/__											
__/__											
__/__											
	Totals this page:										I certify that the entries in this log are true.
	Totals brought forward:										
	Totals to date:										

118

Date	Aircraft	Location		Flight Duration				Function		Ld	Remarks and Endorsements
Yr	Make/Model	Takeoff	Landing	Fixed-wing	Rotary	LTA	Total	OIC/Solo	Student/Instructor	No.	
Dd/ Mm	Registration	Time	Time								
__/__											
__/__											
__/__											
__/__											
__/__											
__/__											
__/__											
__/__											
__/__											
Totals this page:											I certify that the entries in this log are true.
Totals brought forward:											
Totals to date:											

119

Date	Aircraft		Location		Flight Duration				Function			LD	Remarks and Endorsements
Yr	Make/Model		Takeoff	Landing	Fixed-wing	Rotary	LTA	Total	OIC/Solo	Student/Instructor		No.	
Dd/ Mm	Registration		Time	Time									
__/__													
__/__													
__/__													
__/__													
__/__													
__/__													
__/__													
__/__													
Totals this page:													*I certify that the entries in this log are true.*
Totals brought forward:													
Totals to date:													

Date	Aircraft	Location		Flight Duration				Function		Ld	Remarks and Endorsements
Yr	Make/Model	Takeoff	Landing	Fixed-wing	Rotary	LTA	Total	OIC/Solo	Student/Instructor	No.	
Dd/ Mm	Registration	Time	Time								
__/__											
__/__											
__/__											
__/__											
__/__											
__/__											
__/__											
__/__											
Totals this page:											I certify that the entries in this log are true.
Totals brought forward:											
Totals to date:											

DATE	AIRCRAFT	LOCATION		FLIGHT DURATION				FUNCTION		LD	REMARKS AND ENDORSEMENTS
Yr	Make/Model	Takeoff	Landing	Fixed-wing	Rotary	LTA	Total	OIC/Solo	Student/ Instructor	No.	
Dd/ Mm	Registration	Time	Time								
__/__											
__/__											
__/__											
__/__											
__/__											
__/__											
__/__											
__/__											
		Totals this page:									I certify that the entries in this log are true.
		Totals brought forward:									
		Totals to date:									

DATE	AIRCRAFT	LOCATION		FLIGHT DURATION				FUNCTION		LD	REMARKS AND ENDORSEMENTS
Yr	Make/Model	Takeoff	Landing	Fixed-wing	Rotary	LTA	Total	OIC/Solo	Student/Instructor	No.	
Dd/ Mm	Registration	Time	Time								
__/__											
__/__											
__/__											
__/__											
__/__											
__/__											
__/__											
__/__											
	Totals this page:										*I certify that the entries in this log are true.*
	Totals brought forward:										
	Totals to date:										

123

Date	Aircraft	Location		Flight Duration				Function		Ld	Remarks and Endorsements
Yr	Make/Model	Takeoff	Landing	Fixed-wing	Rotary	LTA	Total	OIC/Solo	Student/ Instructor	No.	
Dd/ Mm	Registration	Time	Time								
__/__											
__/__											
__/__											
__/__											
__/__											
__/__											
__/__											
__/__											
		Totals this page:									*I certify that the entries in this log are true.*
		Totals brought forward:									
		Totals to date:									

DATE	AIRCRAFT		LOCATION		FLIGHT DURATION				FUNCTION		LD	REMARKS AND ENDORSEMENTS
Yr	Make/Model		Takeoff	Landing	Fixed-wing	Rotary	LTA	Total	OIC/Solo	Student/Instructor	No.	
Dd/ Mm	Registration		Time	Time								
__/__												
__/__												
__/__												
__/__												
__/__												
__/__												
__/__												
__/__												
	Totals this page:											I certify that the entries in this log are true.
	Totals brought forward:											
	Totals to date:											

125

DATE	AIRCRAFT	LOCATION		FLIGHT DURATION				FUNCTION		LD	REMARKS AND ENDORSEMENTS
Yr	Make/Model	Takeoff	Landing	Fixed-wing	Rotary	LTA	Total	OIC/Solo	Student/Instructor	No.	
Dd/ Mm	Registration	Time	Time								
__ /__											
__ /__											
__ /__											
__ /__											
__ /__											
__ /__											
__ /__											
__ /__											
Totals this page:											I certify that the entries in this log are true.
Totals brought forward:											
Totals to date:											

Date	Aircraft		Location		Flight Duration				Function		Ld	Remarks and Endorsements
Yr	Make/Model		Takeoff	Landing	Fixed-wing	Rotary	LTA	Total	OIC/Solo	Student/ Instructor	No.	
Dd/ Mm	Registration		Time	Time								
__/__												
__/__												
__/__												
__/__												
__/__												
__/__												
__/__												
__/__												
	Totals this page:											*I certify that the entries in this log are true.*
	Totals brought forward:											
	Totals to date:											

127

Date	Aircraft	Location		Flight Duration				Function		Ld	Remarks and Endorsements
Yr ___	Make/Model	Takeoff	Landing	Fixed-wing	Rotary	LTA	Total	OIC/Solo	Student/ Instructor	No.	
Dd/ Mm	Registration	Time	Time								
___/___											
___/___											
___/___											
___/___											
___/___											
___/___											
___/___											
___/___											
		Totals this page:									I certify that the entries in this log are true.
		Totals brought forward:									
		Totals to date:									

128

DATE	AIRCRAFT	LOCATION		FLIGHT DURATION				FUNCTION		LD	REMARKS AND ENDORSEMENTS
Yr	Make/Model	Takeoff	Landing	Fixed-wing	Rotary	LTA	Total	OIC/Solo	Student/Instructor	No.	
Dd/ Mm	Registration	Time	Time								
__/__											
__/__											
__/__											
__/__											
__/__											
__/__											
__/__											
__/__											
Totals this page:											*I certify that the entries in this log are true.*
Totals brought forward:											
Totals to date:											

129

DATE	AIRCRAFT	LOCATION		FLIGHT DURATION				FUNCTION		LD	REMARKS AND ENDORSEMENTS
Yr	Make/Model	Takeoff	Landing	Fixed-wing	Rotary	LTA	Total	OIC/Solo	Student/ Instructor	No.	
Dd/ Mm	Registration	Time	Time								
__/__											
__/__											
__/__											
__/__											
__/__											
__/__											
__/__											
__/__											
		Totals this page:									*I certify that the entries in this log are true.*
		Totals brought forward:									
		Totals to date:									

130

DATE	AIRCRAFT	LOCATION		FLIGHT DURATION				FUNCTION		LD	REMARKS AND ENDORSEMENTS
Yr	Make/Model	Takeoff	Landing	Fixed-wing	Rotary	LTA	Total	OIC/Solo	Student/Instructor	No.	
Dd/ Mm	Registration	Time	Time								
__/__											
__/__											
__/__											
__/__											
__/__											
__/__											
__/__											
__/__											
		Totals this page:									I certify that the entries in this log are true.
		Totals brought forward:									
		Totals to date:									

FLIGHT AND GROUND TRAINING RECORD

Date	Course	Duration	Test Score	Pass Y / N	Instructor / Examiner Name, Signature, Certificate Number

FLIGHT AND GROUND TRAINING RECORD

Date	Course	Duration	Test Score	Pass Y / N	Instructor / Examiner Name, Signature, Certificate Number

INITIAL AND RECURRENT TRAINING

I certify that _____
has satisfactorily completed the aviation training course/
program titled

_____

Signed _____ Date _____

Cert. # _____ Expiration _____

I certify that _____
has satisfactorily completed the aviation training course/
program titled

_____

Signed _____ Date _____

Cert. # _____ Expiration _____

I certify that _____
has satisfactorily completed the aviation training course/
program titled

_____

Signed _____ Date _____

Cert. # _____ Expiration _____

I certify that _____
has satisfactorily completed the aviation training course/
program titled

_____

Signed _____ Date _____

Cert. # _____ Expiration _____

INITIAL AND RECURRENT TRAINING

I certify that _____
has satisfactorily completed the aviation training course/
program titled

_____

Signed _____ Date _____

Cert. # _____ Expiration _____

I certify that _____
has satisfactorily completed the aviation training course/
program titled

_____

Signed _____ Date _____

Cert. # _____ Expiration _____

I certify that _____
has satisfactorily completed the aviation training course/
program titled

_____

Signed _____ Date _____

Cert. # _____ Expiration _____

I certify that _____
has satisfactorily completed the aviation training course/
program titled

_____

Signed _____ Date _____

Cert. # _____ Expiration _____

INITIAL AND RECURRENT TRAINING

I certify that _____
has satisfactorily completed the aviation training course/
program titled

_____

Signed _____ Date _____

Cert. # _____ Expiration _____

I certify that _____
has satisfactorily completed the aviation training course/
program titled

_____

Signed _____ Date _____

Cert. # _____ Expiration _____

I certify that _____
has satisfactorily completed the aviation training course/
program titled

_____

Signed _____ Date _____

Cert. # _____ Expiration _____

I certify that _____
has satisfactorily completed the aviation training course/
program titled

_____

Signed _____ Date _____

Cert. # _____ Expiration _____

INITIAL AND RECURRENT TRAINING

I certify that _____
has satisfactorily completed the aviation training course/
program titled

_____

Signed _____ Date _____

Cert. # _____ Expiration _____

I certify that _____
has satisfactorily completed the aviation training course/
program titled

_____

Signed _____ Date _____

Cert. # _____ Expiration _____

I certify that _____
has satisfactorily completed the aviation training course/
program titled

_____

Signed _____ Date _____

Cert. # _____ Expiration _____

I certify that _____
has satisfactorily completed the aviation training course/
program titled

_____

Signed _____ Date _____

Cert. # _____ Expiration _____

Printed in the USA
CPSIA information can be obtained
at www.ICGtesting.com
LVHW051916031123
762909LV00005B/64

9 783033 058088